Part of the Millennium Wall at Wirksworth, Derbyshire
Artist's impression by David Griffiths

Produced and published by the
Dry Stone Walling Association of Great Britain
Lane Farm, Crooklands, Milnthorpe, Cumbria, LA7 7NH

First Published 2002
Reprinted 2020 on FSC paper by Stramongate Press, Kendal

ISBN (10 digits): 0 9512306 6 2
ISBN (13 digits): 978 0 95 123066 4

The Dry Stone Walling Association is a registered charity and all proceeds from the sale of this book are used to further the objectives of the organisation.

Registered Charity Number 289678

Her Grace, The Duchess of Devonshire (Patron of DSWA) in deep conversation with DSWA President Bryan Hough, right, and Chairman Paul Webley, left, at the Lancashire wall where Ken Clayton is working.

Dry Stone Walls
The National Collection

The story of the Millennium Wall –
an open-air museum of the stones and styles of
dry stone walling that shape our countryside.

Dedication

This book is dedicated to those members of the
Dry Stone Walling Association of Great Britain,
and their supporters, without whose determination
the Millennium Wall Project could not have been achieved.

Contents

Sponsors

Calouste Gulbenkian Foundation
P F Charitable Trust
C D Stephens
Ernest Cook Trust
W H & S Davies: Gwrhyd Specialist Stone Quarry
Duke & Duchess of Devonshire/Chatsworth Country Fair 2000
Northumbria Branch: Guess Weight of the Stone
Buxton Mineral Water Co Ltd
J Riddick
J R Bown
K&H Haulage

In addition to those individuals mentioned in the sections of this booklet, many members made contributions to the success of this project, including:

DSWA Project Team:
John Bown, Brian Jones, Jacqui Simkins

Project Weekend Support:
Bryan Hough, John Riddick, Alison Shaw, Ann Stewart, Paul Webley

Site Works (installation of stone work for signs, etc):
John Bown, Shaun Graney, Gordon Wilton, Jason Wilton, Trevor Wragg

Foreword

I am delighted to be able to offer a few words to introduce this publication.

The Dry Stone Walling Association's Millennium Wall Project was developed by John Bown of Ashover, Derbyshire, who spent eighteen months working to bring his idea to fruition.

The Millennium Wall, showing many of the typical vernacular walling styles of the British Isles, is the only one of its type known to exist. The voluntary efforts of members of the Association from throughout these islands have resulted in a major attraction for the National Stone Centre. The building project was featured on the BBC's "Countryfile" and BBC Scotland's "Landward" television programmes and was the subject of a video made by Alpha Audio.

We hope that our efforts will provide visitors with a better understanding of the range of walling styles found around the British Isles and their relationships to the types of stone from which they are built. This book will be of interest to those who visit the wall and to everyone with an interest in stone or in dry stone walling and we trust it will give you an appreciation of the skills involved in this ancient craft.

Paul Webley
Chairman
Dry Stone Walling Association of Great Britain

March 2002

The Millennium Wall Project

The Vision
A national educational display of the craft of dry stone walling, based on the distinctive styles which are found over all of Great Britain, was an ambition of some members of the Dry Stone Walling Association. Lengths of wall built in local style by local craftsmen, using their own local stone, would demonstrate the skills of the craft and show how walls vary depending on the type of stone available.

The Millennium was the occasion that appealed to the membership for a major project.

The National Stone Centre is located in a region that has an abundance of stone walls and is within easy reach of a large population. The site is ideal for a permanent display of dry stone walls which is always open to the public.

Arrangements to Achieve the Vision
National sponsorship was sought and found for the permanent display and interpretation boards, and for some of the general costs. The site was prepared and arrangements made to accommodate all the participants and the public for the three-day event.

Across the country arrangements were made for the 19 sections of wall, each 6 metres long. Each DSWA Branch, or individual member, had to find local stone (about 10 tonnes), local sponsorship and transportation to the site where stone was stored prior to the event.

The Event
May Day Bank Holiday (29, 30 April and 1 May, 2000) saw nineteen teams of wallers and dykers converging on the National Stone Centre near Matlock in Derbyshire; one from Orkney made a round trip of 1100 miles.

Over 100 people were actively involved: helpers provided food and drink, and produced and manned displays for the public.

The visiting public and groups of Cub Scouts were informed about dry stone walls and watched the Millennium Wall being built.

Members, previously often unknown to each other, enjoyed meeting and discussing the variety of ways walls and dykes can be built.

Information Plinth

The plinth is an unusual structure specially designed to carry the main information panel at the top of the site. It has been made from the local sandstone and limestone with contrasting colours to give an interesting pattern. The outer faces of the stones have been dressed to give a smooth surface and the sides of the stones shaped to that they taper inwards to fit well together. The top is sloping although the courses are horizontal. Because of its small size it has some internal mortar to give it strength.

Stone: Carboniferous Limestone and Sandstone
Wallers: Steven Allen and Trevor Wragg
Sponsors: Stancliffe Stone Company Ltd

Demonstration Wall

This section of wall, in stone from north east Derbyshire, shows the main features of a standard double wall, which is the most common type. It is racked back to show the construction features described later in this book.

Look for the large foundation stones, the through stones, the double-wall construction, the filling and the top stones. Notice how the stones are laid in courses with each stone crossing the joint between the two below. The stones are placed length ways into the wall to give greatest strength.

Stone: Magnesian limestone
Waller: Kevin Shaw
Sponsors: Tarmac Central Ltd, Bolsover Moor Quarry

West Yorkshire

The West Yorkshire wall is a classic double, coursed sandstone wall built in the grand style suitable for an estate wall, rather than a field wall.

The stones have been carefully dressed or sawn to provide very uniform courses. The top stones are dressed to a rounded, semi-circular shape. Large slabs forming the stile go right through the wall for balance and there is a good stone on the ground to prevent erosion.

Stone: Millstone Grit: Sandstone (Carboniferous)
Wallers: Frank Dickin, Ken France, Beverley Howe, John Billington, Adrian Kenny, Roger Depledge, Margaret Ribchester and Laurence Jones
Sponsors: Johnson Wellfield Quarries Ltd

15

Cotswolds

From the Cotswolds is a double, coursed wall. The familiar honey-coloured limestone breaks into fairly regular sheets, although less regular than sandstone, so that a well-coursed wall can be produced. Stones are usually laid tilted slightly outwards to help keep the wall dry as oolitic limestone is subject to rapid chemical weathering when wet.

Less regular stones have been used for the top stones to produce an irregular profile; in a farming situation this will deter sheep from jumping over.

Stone: Oolitic Limestone (Jurassic)
Wallers: Peggy Burke, David Walmesley-Cotham, Ian McEwan, Vincent Gill, Cliff Cooper and Jamie McColm
Sponsors: Natural Stone Market; Wymark Co; Cotswold Frame Makers

South Yorkshire

Double, coursed wall from South Yorkshire is made from a collection of recycled stone. Note the very large foundation stones and also the large slab over the hole, which is supported by strong stones that form the sides.

The hole is a sheep smoot, also called a lunky or hogg hole, built to enable sheep to pass through the wall. In the field these can be blocked by a large stone.

Stone: Gritstone: Sandstone (Upper Carboniferous)
Wallers: John and Sandra Lackenby, Carl Glaves,
                Tom Valentine, Phil Rowell,
                Hilary Richardson and George Wright
Sponsors: Harris Quarries Ltd.

19

South East Scotland

A random, double dyke from South East Scotland. Dyke is the Scottish term for a dry stone wall. This hard igneous stone breaks into very irregular shapes and cannot be easily dressed with a hammer. The wall is therefore random rather than coursed. A cover band provides a surface to accommodate irregular top stones used to protect the wall.

Compare the style and craftsmanship with the West of Scotland dyke, which is made from very similar stone.

Stone: Dolerite or Whinstone: Igneous (Carboniferous)
Dykers: Bruce Curtis and Richard Love
Sponsors: Aggregate Industries UK Ltd

Derbyshire

Double wall, random for the limestone and coursed for the gritstone, built of stones which are local to the Millennium Wall. The limestone breaks into irregular layers so that regular coursing is not easy. Similarly it does not break cleanly across the layers so the joints between the stones are wider.

The squeeze stile between the two sections has "tied" edges with longer stones across the end and running into the face of the wall on alternate layers.

Stone: Carboniferous Limestone and Carboniferous Gritstone (Sandstone)
Wallers: Gordon and Jason Wilton (Limestone); Tony Martin and Shaun Graney (Gritstone)
Sponsors: P.J. Mycock, Friden Grange Farm (Limestone) D.G. Brailsford, Honeycroft Farm (Gritstone)

22

South Wales

A double, coursed wall typical of South Wales. This Blue Pennant sandstone has a different appearance to the other sandstone walls on display. It has natural colouring, due to iron oxide coating older exposed surfaces. It splits well along the bedding planes and also across to produce a uniformly coursed wall with tight joints.

The coursing is horizontal to enable the height of the wall to be constant along the slope of the ground. Note the end of the wall where the stones have been well-laid to bond both across and into the face of the wall.

Stone: Blue Pennant sandstone (Upper Carboniferous)
Wallers: Ken Young and Mike Hall
Sponsors: Pioneer Aggregates (UK) Ltd.

Caithness

From Caithness, a double, coursed dyke. This sandstone forms thin, uniform layers rather like a slate but it is not a true slate since it has not been metamorphosed under high temperature and pressure.

This stone can form a very strong dyke but takes a long time to build since there are many courses. The stones here have been carefully selected to give a decreasing course thickness with height and regular top stones.

Stone: Caithness Flagstone (Middle Devonian)
Dykers: George Gunn and Sue Rainbow
Sponsors: J.W. Sutherland, Caithness Stone Industries Ltd.

West of Scotland

A random, double dyke from the West of Scotland. This hard igneous rock breaks into irregular pieces and cannot easily be shaped by a hammer. Some attempt has been made to form courses to produce a strong wall. The throughstones project and the top stones have been placed into the wall to give a uniform top line.

Compare this with the South East Scotland dyke, which is built with similar stone.

Stone: Quartz Dolerite (Carboniferous)
Dykers: Hugh and Rosie Allan, John Harper, Irwin Campbell and Paul Millard
Sponsors: Tradstocks Ltd., Stirling

South West Scotland

Single and double boulder dyke to demonstrate the Galloway style from South West Scotland. The granite boulders are very hard and have been rounded as a result of being transported in glaciers. They have to be used as they are found, and not moved far. The end parts are single dykes with one stone spanning the width of the dyke. Light can be seen through the gaps, which is not the case with double dykes. The central section has a double dyke supporting a single dyke. This style can, with specialist skills, be built more quickly than dykes using smaller stone.

Stone: Glacial Erratic Granite Boulders (Devonian)
Dykers: Garth Heinrich and Roger Lewis with
              Paul Craven
Sponsors: John Miller, Pibble Forest, Gatehouse of Fleet

Central Scotland

A double, coursed dyke from Central Scotland. This sandstone is less regular so it is not easy to build in regular courses.

The stone has been recycled from a redundant dyke.

Stone: Sandstone (Lower Devonian)
Dykers: Syd Mitchell, John Fenwick, Joyce Anderson,
                Dorothy Spencer, John Cameron,
                Kate Armstrong, May Hirst and
                Margaret Williamson
Sponsors: Strathmore Estates, Glamis Castle;
                Marley Building Materials, Glasgow;
                Keyline Builders Merchants, Dundee.

Isle of Skye

A random, double dyke typical of the Isle of Skye. This hard igneous basalt forms irregular stones, which cannot be shaped. The stones are therefore laid randomly, rather than coursed. This is a recycled field dyke.

Since the walls on the island are at low elevation with a mild, damp climate, and flat stones are rare, a double layer of turf can be used for the top, or cope, to keep the other stones in place.

Stone: Basalt (Tertiary)
Dykers: Hector Nicolson, Neil Tonagh, Louise Kerr and Martin Wildgoose
Sponsors: Geologists' Association;
Skye Transport;
Skye and Lochalsh Enterprise;
John Muir Trust

Cumbria

A double, coursed wall from Cumbria. The slate forms well-defined layers and the stone is easy to split. However, it does not break across the layers with perpendicular edges and the joints between the stones can therefore be wide. The throughstones protrude a little. Top stones are sloping in the opposite direction to the lie of the land so that they hold together, even if one is dislodged.

Stone: Borrowdale Volcanic Green Slate (Ordovician)
Wallers: John Stoddart, Andrew Loudon and Steven Allen with Brian Jones and Mike Houston
Sponsors: Burlington Slate Ltd.
Kirkstone Quarries Ltd.

Northumbria

From Northumbria, a double coursed wall. This hard ganister sandstone forms good uniform layers with varying thicknesses. The step-stile utilises good flags, which pass right through the wall and are balanced. To prevent rocking, they need a good base and sufficient weight of stone on top of them. The top stones have been dressed to a semi-circle.

Stone: Ganister Sandstone (Upper Carboniferous)
Wallers: Gerry Dale, Geoff Mallaburn, James Mason, Ian Muse and Robert Shann
Sponsors: Northern Rock Foundation; Fergusons Transport (Blyth); Scott Bros (Harthope Top Quarry)

Cheshire

A double, coursed wall as found in Cheshire. This is a classic sandstone wall with even coursing and tight joints. Note how the wall has been built with the slope of the ground, displaying both sloping courses and tapered courses so that the stones are not at too great an angle and will not slide. The top stones have been shaped to produce a regular profile.

Stone: Milnrow Sandstone (Carboniferous)
Wallers: Karl Pollitt, Philip Davies and Bernard Hannett
Sponsors: Mrs Doreen Earl, A.M. & D. Earl Quarries;
             NYNAS UKAB.

Lancashire

A double, coursed wall in the Lancashire style. This is a classic sandstone wall, well coursed and with graded courses (note the deeper stones at the base of the wall compared to the depth of those near the top). The throughstones protrude here much more than in most styles. The top stones have been trimmed to form a uniform profile.

Stone: Haslingden Flags: Millstone Grit (Carboniferous)
Wallers: Brenda Koo, Carl Watson, Eddie Thurrell and Kenneth Clayton
Sponsors: Aggregate Industries UK Ltd

Sutherland

A Sutherland single boulder dyke made from hard quartzite.
These large boulders occur naturally from the fracturing
of larger boulders or from outcrops. They are angular and
cannot be shaped by a hammer so they have to be used as
they are. The dyke is a single stone in width. Look at the
careful balancing and wedging that hold it together. Note
the dyke ends particularly, since these must be very strong.
The top of the dyke has been made uniform.

Stone: Welded Quarzite (Cambrian)
Dykers: Dave Goulder with Mary Gilchrist
               and Don Eland
Sponsors: Aggregate Industries UK Ltd., Ardgay

Cumbria

Cumbrian Brathay Wall; a slate slab wall found where the slate occurs in large, strong sheets. These are found in the southern Lake District, north Wales and south Lancashire. About a third of the slab is buried for support and the edges are chamfered to interlock. In some regions wires are used to hold the sheets together. Although these are sheep walls they are not very high.

Stone: Brathay Blue Slate (Silurian)
Wallers: Andrew Loudon and John Stoddart
Sponsors: Burlington Slate Ltd and
              Kirkstone Quarries Ltd.

North Wales

North Wales clawdd; this is a Welsh term for wall. This stone-faced earth bank is formed of hard, igneous and granite boulders rounded by the action of glaciers, sea or streams. The stones are wedged together before the earth is added. These types of wall are also found round the west coast of Wales, England and Scotland. In the mild coastal climate the grass top thrives and the wall becomes well vegetated. It is not high since it is used mainly for cattle.

Stone: Glacial Boulder Clay and Moraine Boulders
Wallers: Sean Adcock, Brenda Lewis, Ivan and
               Kathryn Hewitt, Bryn Lewis, Don Eland
               and Gareth Pritchard
Sponsors: Tarmac Quarry Products

A brief history of the craft in Britain

Dry stone walling in Britain stretches back at least three and a half millennia, to the village of Skara Brae in the Orkneys, and the Iron Age brochs of northern and western Scotland. Dry stone walls are found mainly in upland areas of Britain and elsewhere, where large quantities of rock and stone are found near the surface and where trees and hedges do not grow easily because of the climate, elevation, strong winds or thin soils.

The earliest field walls were built of stones cleared from the adjacent land, so that the ground could be cultivated more easily, the farm animals kept safe and shelter from the wind and rain provided for the animals and crops. Quarrying stone for walls came later. Dating walls is notoriously difficult although it is sometimes possible to identify on a site the changes in style and to deduce which is the oldest by the way one may have been built over another. However, it is difficult to tell if the walls have been rebuilt over time.

It is likely that the walls of small rounded fields in parts of Cornwall date back two or even three millennia. These are made from the fieldstone, which is rounded, so that the walls are not easy to build. They are therefore low and have turf or soil between the stones to hold them together.

The rounded fields need less length of wall, and therefore less effort, for the same field space. As farms became more prosperous, larger fields were created and the smaller ones near the houses were used to keep the animals at night.

Ancient Cistercian walls in Derbyshire

Anglo-Saxon and Scandinavian settlement in the north of England led to the extension of fields, especially to the development of the open-field system. These communal fields were fenced off from the common grazing or 'waste'. Linton-in-Wharfedale and Langdale-in-Westmorland have walls dating at least from Norman times.

The walling of smaller fields reached a height in the Elizabethan period when cottagers and householders were, for the first time, legally permitted to enclose small 'crofts' or private holdings. The patterns of small fields around many Pennine villages date from this period. This creation of more fields continued piecemeal, for 200 years with the fields becoming larger, the walls straighter and with squarer corners, as the population grew and the open-field system broke down.

Mixed stone types and sizes in a wall as woud be found in the Langwathby area of Cumbria

Walling changed with the large-scale enclosures from about 1780, promoted by landowners or entrepreneurs who could engineer private Acts of Parliament to abolish common rights. Teams of professional wallers and dykers appear, hired to build many miles of walls quickly. Exacting specifications survive from this period, and many walls still bear evidence of their origin, with precisely placed throughstones and topstones, uniform batter, and

unvarying height. The walls are usually straight and do not deviate to incorporate large rocks or to avoid wet areas. In the Pennines this enclosure movement was finished by about 1820; in the Lake District by the end of the 19th century. Organised Scottish enclosure had begun in the early 1700s. By the end of the nineteenth century there were many estate walls, often of high quality.

The 'enclosure style' is now the accepted high standard that is encouraged by the Dry Stone Walling Association but walls should always reflect the nature of the local stone and local vernacular style.

By 1900 there were few areas left to be sub-divided so that most walls were modified, rebuilt and repaired by farm workers until the drastic fall in the numbers of staff on each farm, which continued throughout the 20th century. The heavy industrialisation of farming between 1950 and 1980 produced a decrease in the appreciation of the long-term value of walls and was the low point of the craft in Britain.

Today dry stone walling is prospering with an upsurge in interest in the environmental value of walls and with the growth in prestigious garden, landscape and artistic projects adding to the range of work available to professional wallers and dykers.

What is a dry stone wall?

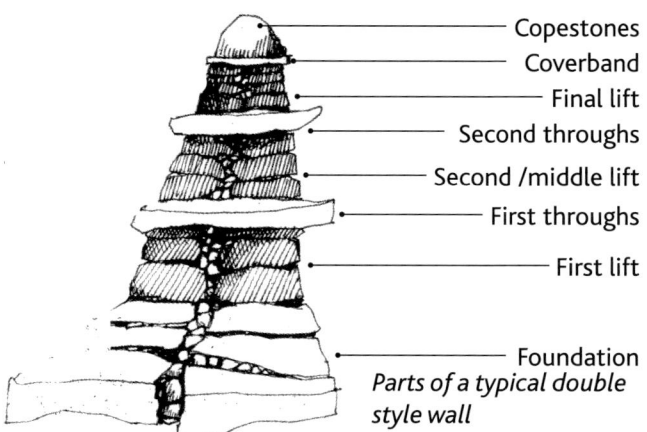

- Copestones
- Coverband
- Final lift
- Second throughs
- Second /middle lift
- First throughs
- First lift
- Foundation

Parts of a typical double style wall

Dry stone walls are a practical form of farm field boundary. Although initially more costly to build, a dry stone wall will outlast a wire fence many times over and, given timely maintenance, will prove more economical in the long term. Dry stone walls enhance the landscape, are part of our heritage and are found in many parts of Great Britain, and overseas, wherever field, mountain or quarried stone is plentiful. They are used to keep farm and wild animals in their place, to protect growing crops and to provide shelter. They are also used in the simpler forms of barn or house building and as features.

Dry stone walls are built without mortar or cement, reducing the cost and requiring no materials not readily available. More importantly, the flexibility of the joints allows the wall to survive frosts and to settle safely if the underlying ground is slightly soft.

The essential wall features

Walls can be built from most types of stone: the skill is in making the best use of what is available. The picture shows a section of the most common type of wall. This is a double wall with throughs. We will use this to describe its main features and contrast these with those of the walls that you will see in the Millennium Wall. The glossary on page 61 may also help.

At the base are some large flat stones, the foundations, to spread the considerable weight (about a tonne a metre) over a large area of the ground. These may cover ground wider than the wall itself if the ground is soft. On this are built two walls, with good outer faces and smaller stones, or hearting, carefully placed between them to prevent the stones sliding inwards. Large throughstones are laid across the two outer walls at about half the height to tie them together. Then there is another lift of doubled wall with perhaps a further course of throughstones. To finish, there are top, or cope stones, which again bind the two faces together and provide more height. As you will see in the Millennium Wall, there are regional variations, but the same basic principles of construction apply.

If the geology of the stones is such that they form slabs with uniform thickness, they are best laid in courses. If the stones are angular and have an irregular shape with few flat surfaces then they are laid in a random arrangement. Very large stones are difficult to move and a double wall is unnecessary so only a single thickness of stone is used. If they are made of a hard stone which has had all its corners ground away by the action of the sea, a stream or a prehistoric glacier to form a "ball" shaped stone, then special techniques are necessary to make a strong wall.

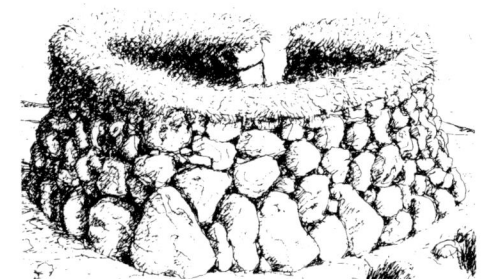

Boulders used to construct a "single" dyke

Quality of a wall

The best test of the quality of a wall is how long it stands up: but this may take a long time to determine. However, the skill of the craftsman can be seen immediately, especially while the wall is being built. The faces of the wall should be uniform and smooth with a constant and gentle batter, or taper so that the wall becomes narrower with height. If there are courses, then each should be of uniform height and the stone should become thinner towards the top. Whether the wall is coursed or random, the top of the wall below the cope stones should be level and the top of the cope stones should also be regular and at a constant height from the ground.

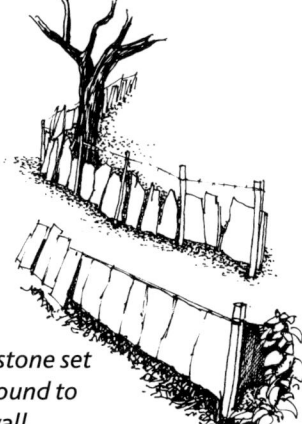

Flags of stone set in the ground to form a wall

The throughstones should not protrude much beyond the face of the wall but should look similar and be placed at the same height in the wall. The hearting should be well placed and sufficient so that the bigger stones do not move. The building stones should touch each other as much as possible so that the gaps appear narrow. All stones must be held firmly in place so that none can be pulled out or moved easily.

Building a dry stone wall

Individuals and groups who are interested in learning more about the practical craft of dry stone walling are encouraged to participate in a training course. The DSWA branch network offers weekend courses suitable for beginners.

It is important to remember that stone is not uniform. The following drawings demonstrate construction of the commonly occurring doubled wall, using level-bedded stone, but exactly the same process is required for irregular stone although the appearance will differ. Walls are usually 1.4m high, built of a base called a foundation or footing, 75-80cm in width.

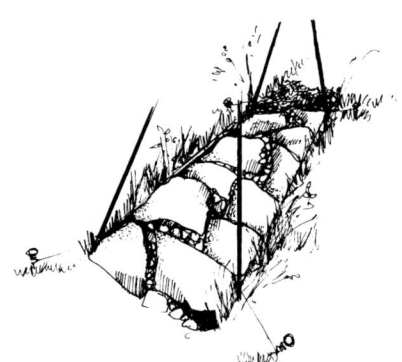

Foundation stones laid in trench, packed with smaller stones between them.

The foundations will carry all the weight of the wall and therefore the largest stones with the greatest surface area should be used.

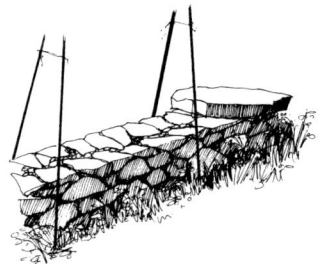

First lift levelled off and one large throughstone in position.

Each course consists of two "walls" carefully filled with "hearting". The golden rule for laying stones is one stone on two, and two stones on one, ensuring every joint is covered by the course above; another is to lay stones lengthways into the wall, resisting the temptation to make speed by laying the length along the face of the wall. Each piece of hearting should be carefully placed to pack the stones firmly: it is never piled in.

As the building proceeds, each course is slightly narrower than that immediately beneath it: this produces the "batter" on the wall – the gently tapering sides.

At strategic levels in the wall, usually about half way up the height, throughstones are placed, which tie the two faces together. In some regions of Britain the throughstones protrude beyond the face of the wall. Occasionally, where plentiful, a continuous row of throughstones is laid in a wall.

To finish the top of the wall are cope stones, sometimes set immediately on the "double", sometimes with a coverband (final course of stones laid the full width of the wall) directly beneath them. Cope stone styles vary greatly throughout Britain: some are set upright, some at 45 degrees; some are trimmed to equal height, others are deliberately set with a variety of heights. Occasionally they are set flat, as in the coverband, and in some regions tall and short stones are set alternately. These variations are localised and it is important to maintain the vernacular styles.

When watching a waller or dyker at work, you might notice the use of a walling frame and strings to keep the wall straight and ensure the batter remains even as the wall progresses (see drawings on previous page). The waller will make little use of a hammer when repairing field walls, although landscape or garden work will require more shaping of stone to provide the desired finish and artistic blend of stone to suit the designs.

Dry stone walling includes many features sometimes called furniture, such as the lunky, or hogg hole or smoot shown above: note the long lintel stone. These features have many local and occasionally unusual names.

Stone for building walls

People frequently enquire where the stone came from for the walls in the countryside. The ancient field walls in the landscape used stone from as near to hand as possible. Sometimes this was "field stone", pieces that came to the surface as the fields were ploughed, and the walls formed with this stone often enclosed relatively small areas beside the homesteads; or they became the wide-based "consumption" or "accretion" walls which slowly got wider as more stone that came to the surface with deeper ploughing was removed to the edge of the field and added to the wall.

Stone for many upland walls, especially Enclosure Act walls, came from shallow pits or quarries close to the building work. The shallow quarries can sometimes still be seen as grassy hollows beside the wall, often quite short distances apart. The stone was hewn by hand, and moved to where it was required: it would have been moved down the slopes – gravity helping movement. It is thought that most was moved by manpower but perhaps with the help of a simple sled pulled by a man or maybe occasionally by a horse.

Stone required for enclosing formal grounds around manor houses was sometimes local, but more often than not imported from elsewhere. Cottages and farmhouses were built of local stone, and what remained after completion of the house and outbuildings would be used to enclose the yard and garden, providing shelter from weather for crops and produce, whilst also excluding farm stock.

Stone for repairs to field walls frequently comes from adjacent redundant walls, whilst that for new work is usually from a commercial quarry. Some quarries produce stone for specialist building projects, and that for dry stone walling often utilises stone not suitable for cut or dressed building work. Some quarries use blasting techniques which render stone unsuitable: it contains small fractures which break apart when the stone is struck, and which also permit water to enter the stone, breaking it open when frost occurs.

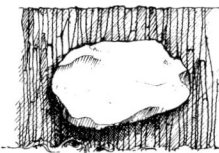

Artistic use of dry stone work: a large boulder supported in a contrasting wall of vertically laid stone.

There are a number of publications on quarries and quarrying, and on stone in general, which will make useful reading for those wishing to pursue their interest.

Wall styles, location and geology

Walls usually occur where there is readily available stone; stone is extrmeley heavy and costly to move. Stone that was lying on or in the top few inches of the ground, or easily pulled out of rocky outcrops higher up the hillside from where the wall was to be built, was the first to be used. Thus walls occur where the soil is thin and rocks are plentiful. The style of the wall is determined by the physical properties of the local stone, which affect the way it can be used.

Sedimentary rocks (sandstones, limestones and mudstones) were formed by the cementing together of particles that were laid down in level beds when settling under water. These particles were originally eroded from the land mass and washed into rivers or the sea, or were formed from the shell fragments of fossil organisms. Even after compacting under pressure, they are able to be worked using these distinct "bedding planes" to split the rock into stone of varying thickness

Cheekend in typical Pennine style

with flattish top and bottom surfaces. These are then built into regularly coursed walls, characteristic of the Pennines area. Some carboniferous limestones, however, do not split easily and are used in a more random manner of building. These can be found in the walls of the Derbyshire Peak District, north and south Wales, Somerset, the Yorkshire Dales and parts of the Lake District.

Metamorphic rocks were formed when sedimentary or igneous rocks were subjected to high temperature and pressure. Their properties have been changed: they become crystallised and sometimes have good cleavage planes. This enables the rocks to be split and used in dry stone walling, as in the upright slate Brathay Wall.

Igneous rocks are crystalline and the products of the magma deep in the earth's crust. Some have cooled slowly at depth, others more rapidly on the surface. They are often very hard, have no bedding planes and have to be used largely as they are found. These hard rocks are built into random walls, boulder dykes, the dykes on Skye and the Welsh cloddiau are examples. Sometimes these rocks have been rounded into large pebbles as a result of the action of rivers or glaciers.

Timescale of Rocks

More than 540 MA (million years ago)	Precambrian	Gneisses, schists, sandstones, conglomerates, siltstones: Hebrides and NW Scotland coast; A few southern outcrops: Anglesey, Charnwood, and Long Mynd.
540 to 500 MA	Cambrian	Shales, slates, gritstones. Harlech Dome, Malverns, North Pembrokeshire, Isle of Man; in Scotland adjacent to Precambrian.
500 to 410 MA	Ordovician, then Silurian	Shales, mudstones, some limestones: Mid-Wales, extending into Pembrokeshire and Denbighshire; central and southern Lake District, South Uplands. "Caledonian Orogeny" or mountain building following continental collision caused some metamorphism south of the Border, but very extensive north of the Highland Line. Granite emplacements and volcanic rocks.
410 to 345 MA	Devonian	Sedimentary rocks and granite emplacements: Cornwall and South Devon: Old Red Sandstone in S Wales, N Devon, Midland Valley Scotland. Moray Firth and Caithness.
345 to 280 MA	Carboniferous	Limestone, Millstone Grit, Coal Measures, Sandstones: S Wales, Pennines, Cumbria, Midland Valley of Scotland. Culm in mid Devon.
280 to 210 MA	Permian and Triassic	Magnesian Limestone, New Red Sandstone, conglomerates: Somerset, Midlands, E & W of Pennines, N & E of Lake District.
210 to 150 MA	Jurassic	Limestones, shales: Dorset to North Yorkshire coasts.
150 to 65 MA	Cretaceous	Greensands, Chalk: South East England, Yorkshire.
65 MA to present	Tertiary	Soft rocks in South East of England; Volcanic rocks in West of Scotland.

DRY STONE WALLING ASSOCIATION

The Dry Stone Walling Association (DSWA) is a registered charity dedicated to promoting the craft, heritage and conservation of dry stone walling. It can provide advice and guidance on all matters relating to this traditional skill and offers training opportunities for different levels of ability. A list of current professional members is produced each year in order to help locate the services of a dry stone waller if required.

As a membership organisation there are numerous ways you can get involved or help with the work we do. For those wishing to practise their skills, the branch network offers a choice of walling projects from taster days and beginner courses through displays at events and shows to community projects and social events. If practical walling projects are not your thing, you can still join the Association and your membership fee will enable us to continue to build on our work highlighting the importance of dry stone walls that are such an iconic part of the landscape, not just as boundaries but as wildlife corridors for small mammals and a habitat for insects and plants. Your support can also help to train young people as "the new generation of wallers" to keep the skills alive for the future. A members' magazine, Waller & Dyker, is produced three times a year and is packed full of articles and information about a wide variety of walling related subjects and there are opportunities to take part in events throughout the country.

Please take a look at our website, www.dswa.org.uk, for the full range of activities, including educational resources and a shop, or contact the office at Lane Farm, Crooklands, Milnthorpe, Cumbria, LA7 7NH. information@dswa.org.uk

The National Stone Centre

Set within six former limestone quarries in the heart of the Derbyshire Dales, on the edge of the Peak District National Park, and close to the Derwent Valley Mills World Heritage Site, the National Stone Centre (NSC) is a 40 acre Site of Special Scientific Interest (SSSI) for its geological formations, offering outdoor and indoor activities for all.

- Outdoor fossil trails around our free to access site
- Visitor Centre with shop, café and "Building Britain" Exhibition
- Geo walks and picnic areas
- Childrens play area

The National Stone Centre also hosts popular introductory courses in dry stone walling and traditional stone carving techniques.

The National Stone Centre can carry out stone sourcing and matching for bridges, buildings, churches, walls and historical research relating to all sectors of the quarrying industry and has a library (accessible by prior appointment) covering geology, the minerals industry, minerals planning and a school resources section. It also has a large collection of geological material and photographs, and a small archive on quarry history. All these are at the stage of initial curatorial work and documentation.

http://www.nationalstonecentre.org.uk

Porter Lane
Middleton by Wirksworth
Matlock
Derbyshire DE4 4LS

A Brief Glossary
Terms commonly used in the craft of dry stone walling and dyking, some of which appear in this booklet.

A-FRAME: a wooden or metal frame used as a guide when building.

BATTER: the inward taper of the wall from base to top.

CLAWDD: Welsh term for wall. Plural is cloddiau.

CONSUMPTION DYKE: wall built with stone cleared from the land and which is especially wide. Also called "clearance wall" and "accretion wall".

COPE STONES: the top stones, the stones along the top of the wall to give weight and protection. Also called "cams", "tops", "toppers".

COURSE: horizontal layer of stones placed on a wall.

COVERBAND: large flat stones placed across width at top of wall in some areas to form base for the cope stones.

DOUBLING OR DOUBLE: term used for a dry stone wall built with two faces of stones, packed with hearting between.

DRESSED: term used to describe the process of shaping stone with a hammer.

DYKE /DYKER: Scottish term for a dry stone wall/person building walls.

FOUNDATION: the first layer of large stones in the base of the wall, also called "footings" or "founds".

GALLOWAY DYKE: wall or dyke with lower third "doubled", upper two thirds in single walling.

GAP OR GAPPING: a breach in a dry stone wall. Gapping is the repair of same and the "gapper" is the waller or dyker who carries out the repair.

HEARTING: the small stones used as filling or packing in a double wall.

PINNINGS/PINS: small, usually tapering stones used from inside the wall to wedge building stones firmly in place.

RETAINING WALL: dry stone wall built into the cut face of a bank to prevent the soil from moving down the slope.

SINGLE DYKE: wall built with single stones going the width of the wall.

THROUGHSTONES: large, heavy stones placed at regular intervals along the wall to tie the two sides together.

TRACE WALLING: incorrect placing of stones with their length along face of wall rather than placing into the wall for strength.

WALLHEAD: vertical end to a length of wall. Also called "cheekend".

Visit the Wall

Start at the sections beneath the trees and then walk up the slope. For best effect, stand back from the wall when observing each section.

REGION	STONE	NOTES
North Wales	Glacial boulders	Stone faced earth bank or "clawdd".
Cumbria	Slate	Flag wall of vertical slabs as found in parts of Lake District
Sutherland	Welded quartzite	Single boulder dyke
Lancashire	Sandstone (gritstone)	Coursed, double wall
Cheshire	Carboniferous sandstone	Coursed, double wall
Northumbria	Sandstone (ganister)	Double wall incorporating step-stile
Cumbria	Burlington slate	Coursed, double wall; sloping copes
Isle of Skye	Basalt laval flow (reclaimed)	Random, double dyke; turf top
Central Scotland	Reclaimed sandstone	Double dyke
South West Scotland	Reclaimed granite	Galloway dyke
West of Scotland	Quartz dolerite	Double dyke
Caithness	Slate	Coursed, double dyke
South Wales	Blue Pennant sandstone	Coursed double wall
Derbyshire	Sandstone (gritstone) & carboniferous limestone	Double wall: squeeze-stile separates stone types. Stone from redundant walls.
South East Scotland	Carboniferous dolerite	Random, double dyke with coverband
South Yorkshire	Gritstone salvage offcuts	Double wall with lunky (sheep hole)
Cotswolds	Jurassic oolitic limestone	Double wall
West Yorkshire	Carboniferous sandstone	Dressed stone; double wall with stile
Plinth	Limestone & gritstone	Local Derbyshire stones
Demonstration Wall	Magnesium limestone	Open ended to show basic wall structure

A Short Bibliography

Building & Repairing Dry Stone Walls
ISBN 0 9512306 2 X
Published by DSWA.
Author R Tufnell.

Dry Stone Walling – Techniques And Traditions
ISBN 0 9512306 8 9
Published by DSWA.

Dry Stone Walling
ISBN 0 9502623 0 7
Published by SW Scotland Branch. Reprinted 1999.
Author: Col F Rainsford Hannay.

Walls In The Landscape - Celebrating the Craft of Dry Stone Walling
ISBN 978 0 9568458 2 5
Published by DSWA

In There Somewhere
ISBN 0 9512306 5 4
Published by DSWA.
Written and illustrated by David Griffiths

Dry Stone Walls
ISBN 978 0 7478062 0 2
Published by Shire.
Author: Lawrence Garner.

The Slate Industry
ISBN 978 0 7478 0124 5
Published by Shire.
Author: Merfyn Williams.

Dry Stone Walling – A Practical Handbook
ISBN 978 0946 752195
Published by The Conservation Volunteers (formerly BTCV)
Revised by Elizabeth Agate with Sean Adcock 1999.

The Hidden Landscape – a journey into the geological past.
ISBN 0-7126-6040-2
Published by Pimlico, 1996.
Author: Richard Fortey

Fields in the English Landscape
ISBN 9780862994495
Published by Sutton Publishing Ltd.
Author: Christopher Taylor

The line drawings used in this publication are
taken from DSWA publications, mostly
In There Somewhere,
written and illustrated by David Griffiths and
published by the Dry Stone Walling Association.

Photographs have been provided by members
and supporters of the Association, for which we
are very grateful.

Front cover background image courtesy of
Janet Matthews – Inishmaan, Aran Islands.